A STEP INTO HISTORY™

THE GREAT DEPRESSION

BY STEVEN OTFINOSKI

Series Editor

Elliott Rebhun, Editor & Publisher,

The New York Times Upfront

at Scholastic

Content Consultant: James Marten, PhD, Professor and Chair,
History Department, Marquette University, Milwaukee, Wisconsin

Cover: Men wear signs asking for work.

Library of Congress Cataloging-in-Publication Data
Names: Otfinoski, Steven, author.
Title: The Great Depression / by Steven Otfinoski.
Description: New York, NY : Children's Press, an imprint of Scholastic Inc., 2018.
| Series: A step into history | Includes bibliographical references and index.
Identifiers: LCCN 2017025779 | ISBN 9780531226902 (library binding) |
ISBN 9780531230121 (pbk.)
Subjects: LCSH: United States—History—1933–1945—Juvenile
literature. | United States—History—1919–1933—Juvenile literature.
| Depressions—1929—United States—Juvenile literature. | New
Deal, 1933–1939—Juvenile literature. | United States—Economic
conditions—1918–1945—Juvenile literature.
Classification: LCC E806 .O78 2018 | DDC 330.9730916—dc23
LC record available at https://lccn.loc.gov/2017025779

 Published in 2018 by Children's Press, an imprint of
Scholastic Inc. Printed in Johor Bahru, Malaysia 108

Scholastic Inc., 557 Broadway, New York, NY 10012.

1 2 3 4 5 6 7 8 9 10 R 27 26 25 24 23 22 21 20 19 18

CONTENTS

PROLOGUE 6

KEY CONCEPTS FOR UNDERSTANDING THE GREAT DEPRESSION 10

MAPS 12

CHAPTER 1 • **SPIRALING SPECULATION** 14

CHAPTER 2 • **BLACK TUESDAY** 18

CHAPTER 3 • **BANKS GO BUST** 22

CHAPTER 4 • **A PRESIDENT FALTERS** 26

CHAPTER 5 • **EKING OUT A LIVING** 30

CHAPTER 6 • **THE BONUS ARMY ON THE MARCH** 34

CHAPTER 7 • **THE SHAME OF A NATION** 38

CHAPTER 8 • **CHILDREN OF THE DEPRESSION** 42

CHAPTER 9 • **RIDING THE RAILS: HOBO LIFE** 46

CHAPTER 10 • **SONGS OF HOPE AND HAPPINESS** 50

CHAPTER 11 • **THE ELECTION OF 1932** 54

CHAPTER 12 • **NEW PRESIDENT, NEW DEAL** 58

CHAPTER 13 • **DAMS, ROADS, AND BRIDGES** 62

CHAPTER 14 • **FIRESIDE CHATS AND BELLY LAUGHS** 66

CHAPTER 15 • **HAPPY DAYS ARE HERE AGAIN** 70

CHAPTER 16 • **LOST IN THE DUST BOWL** 74

CHAPTER 17 • **DEPRESSION WRITERS** 78

CHAPTER 18 • **THE HARDEST HIT: AFRICAN AMERICANS**... 82

CHAPTER 19 • **PUBLIC ENEMY NUMBER ONE** 86

CHAPTER 20 • **THE DEPRESSION IN PICTURES** 90

CHAPTER 21 • **THE RISE OF DICTATORS** 94

CHAPTER 22 • **THE KINGFISH** 98

CHAPTER 23 • **THE SOCIAL SECURITY ACT OF 1935** 102

CHAPTER 24 • **THE ELECTION OF 1936** **106**

CHAPTER 25 • **THE HOLLYWOOD DREAM MACHINE** **110**

CHAPTER 26 • **THE POWER OF ORGANIZED LABOR**......... **114**

CHAPTER 27 • **MURALS ACROSS AMERICA**................ **118**

CHAPTER 28 • **THE WORLD OF TOMORROW**............... **122**

CHAPTER 29 • **A NEW WAR**.................................... **126**

CHAPTER 30 • **THE LEGACY OF A DARK TIME** **130**

KEY PLAYERS **134**

GREAT DEPRESSION TIMELINE **136**

GLOSSARY.. **140**

FIND OUT MORE **142**

INDEX.. **143**

ABOUT THE AUTHOR **144**

PROLOGUE

IN THE 1920S, THE UNITED STATES ENTERED A period of great prosperity. The so-called Roaring Twenties was a dizzying decade of fashion, fads, and fun for some Americans. The rich got richer, and the middle class had money to buy new consumer products such as automobiles and radios.

But the prosperity of the 1920s wasn't all it appeared to be. The agricultural market was in decline, and many manufacturing businesses were beginning to find that they had produced more goods than they could sell. Even some banks were starting to face difficulties. Most people did not notice these signs that the economy was starting to fail. Thanks to the booming stock market, they were optimistic about the economy. But the rising stock market was not built on true financial value, but on stocks bought using loans from banks and other institutions. The market finally fell back to earth in late October 1929, igniting the biggest economic disaster in history.

The Wall Street crash affected everyone, including millions of citizens who had not even invested in the stock market. Many businesses went bust. This put their employees out of work. People without jobs could no longer afford goods and

You will find the definitions of bold words in the glossary on pages 140–41.

Find out more about people whose names appear in orange and bold on pages 134–35.

services. This caused many more businesses to close. In only a few months, the United States went from a place of prosperity and certainty to one of poverty and doubt.

No one seemed to know what to do to fix the problem, not even the president, **Herbert Hoover**. It would take a new president, elected in 1932, and a new way of problem solving to begin rebuilding America. But it didn't happen overnight. For more than a decade—a period known as the Great **Depression**—the people of the United States would face want and worry.

The United States had been through economic depressions before. But none lasted as long or affected as many people as this one. The Great Depression changed America and the entire world in many ways, both good and bad.

A man tries to sell his car after going broke during the Great Depression.

KEY CONCEPTS FOR UNDERSTANDING THE GREAT DEPRESSION

WHAT IS THE STOCK MARKET?

It is a place where shares, or units of ownership, in a company are bought and sold. People buy stocks with the idea of holding them for a while and then selling them when they hopefully go up in value. The difference between what they paid for the stocks and the higher price they sell them for is called profit. If the stock price has gone down by the time they sell it, the difference is called a loss.

WHAT HAPPENS WHEN THE STOCK MARKET CRASHES?

Stocks go up and down in value on a daily basis. A crash happens when the market has a major drop in value and does not recover from it in a reasonable period of time. A crash can wipe out all the money, or investment, a person made in a stock.

WHAT CAUSED THE STOCK MARKET CRASH OF 1929?

There are a number of reasons for the crash. Many people bought stocks on credit, using loans to pay for stocks. When stock prices fell, they couldn't afford to pay back the loans, and they lost everything. Other factors that led to the crash included a weak banking system, a poor agricultural market, and companies manufacturing more goods than they could sell.

WHAT ARE INFLATION AND DEFLATION?

Inflation is an overall increase in the prices of goods and services. Inflation makes things more expensive. Each dollar is worth less. After the stock market crash, inflation fell below zero. That's called deflation, an overall decrease in the prices of goods and services.

WHAT ARE DEPRESSIONS AND RECESSIONS?

A depression is a long-term downturn in economic activity. A recession is a slowdown in economic activity for a relatively short period of time.

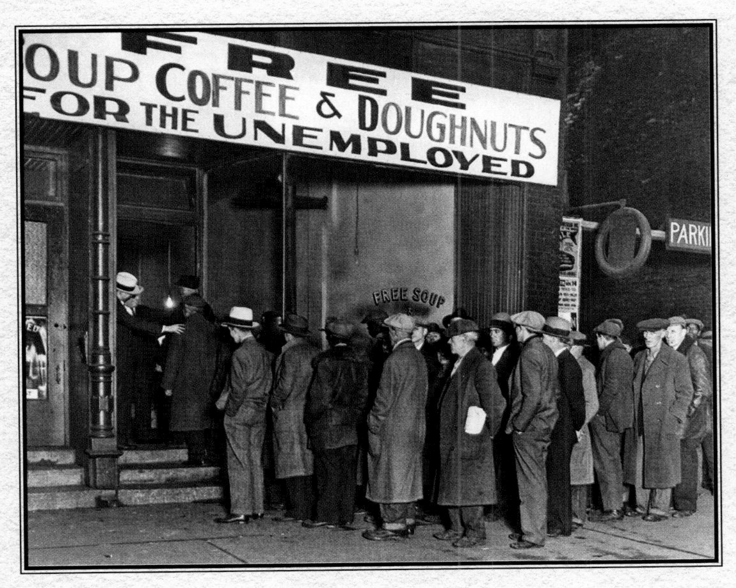

WHY WAS THIS PERIOD CALLED THE GREAT DEPRESSION?

The Great Depression was the longest and most significant depression in American history, lasting from about 1929 to 1939. It affected all aspects of the economy, causing many businesses to cut back drastically or close. Millions of people lost their jobs and homes.

MAPS

KEY LOCATIONS OF THE GREAT DEPRESSION

Major events that helped define the Great Depression occurred in many places across the United States between 1929 and 1939.

THE DUST BOWL

The Dust Bowl was a period during the 1930s when extreme drought and dust storms devastated a huge range of land in America. Thousands of people fled the dry, dusty conditions to California. Many of them traveled hundreds of miles along the famous Route 66 highway.

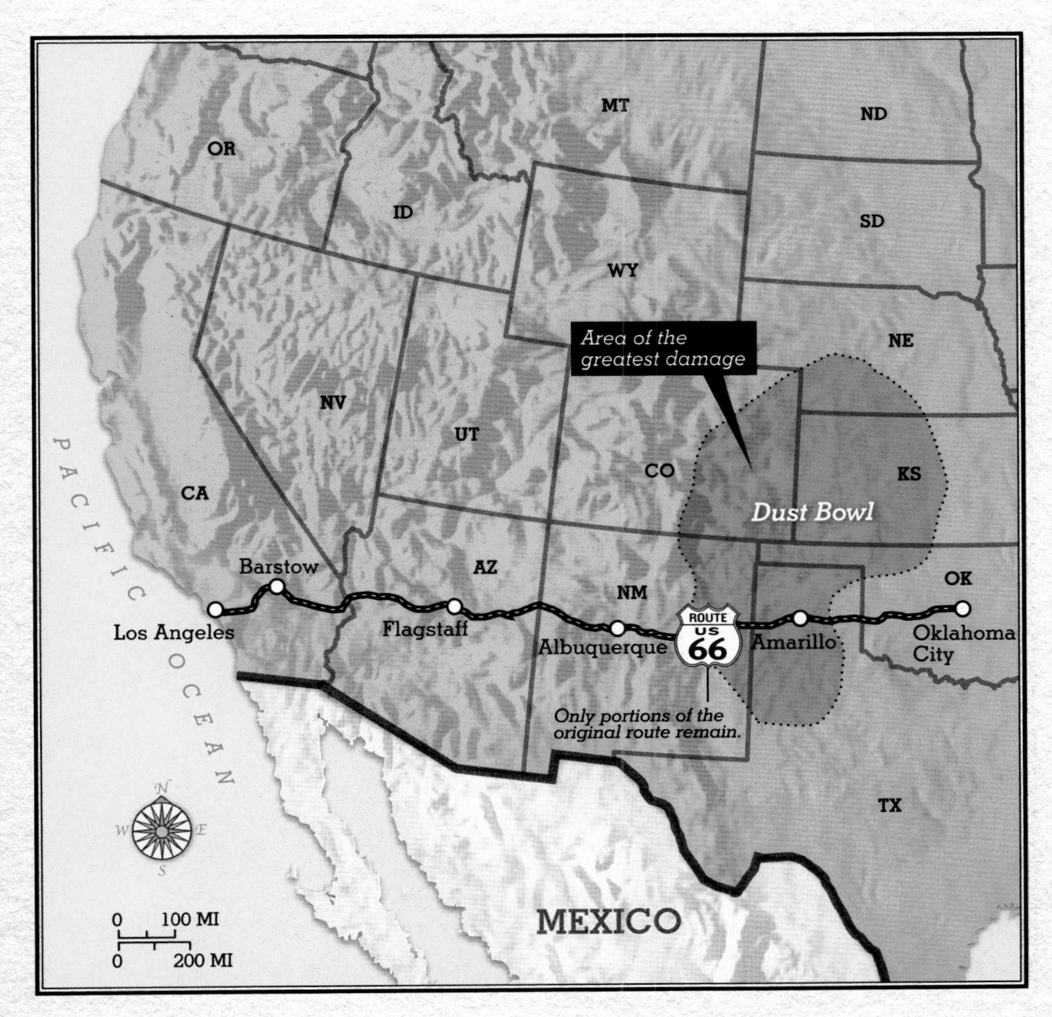

A stock certificate for six shares of the New York Consolidated Railroad

CHAPTER 1

SPIRALING SPECULATION

The Roaring Twenties seemed like an unending party for many Americans, but the good times wouldn't last forever.

THE 1920S SAW ECONOMIC GROWTH IN THE United States unlike anything the nation had experienced before. Factory assembly lines turned out millions of automobiles and other consumer products. Corporate profits rose 62 percent from 1923 to 1929, and the standard of living improved for most Americans. At the same time, workers' wages generally remained low, and the top 5 percent of the population held one-third of the country's total personal wealth.

Many people in the remaining 95 percent pursued another path to riches. For only a few dollars, small investors could buy stocks on margin. This meant they had to put up only a small percentage of the stock's cost, and **stockbrokers** would loan them the rest. The stockbrokers, in turn, got the money to pay for the stocks from bank loans. This kind of **speculation** greatly inflated the price of stocks and would eventually spiral out of control. But everyone was enjoying the ride so much that they refused to think about any of the risks.

A huge crowd gathers in front of the New York Stock Exchange.

NEW YORK STOCK EXCHANGE

"A hotel clerk says to a guest, 'Do you want the room for sleeping or jumping?'"

—A POPULAR JOKE AFTER THE STOCK MARKET CRASH

CHAPTER 2

BLACK TUESDAY

The Great Depression started with a crash that was heard across the nation.

AT ABOUT 11:00 A.M. ON THURSDAY, October 24, 1929, stock prices started to fall, and there was a selling frenzy at the New York Stock Exchange. Brokers frantically called investors and demanded that they pay back the money they owed from buying stocks on margin. When the investors couldn't pay, they lost all their money. The brokers turned to the banks for more loans to make up the difference, but banks didn't have the money either.

A group of wealthy bankers pooled their resources and stopped the slide temporarily. But it was like putting a bandage on a festering wound. Five days later, the bottom fell out of the market. Huge companies teetered on the edge of **bankruptcy**. The day was aptly called Black Tuesday, but there would be darker days to come. By mid-November, money was scarce. Many companies—unable to sell their products and services—closed their doors or fired most of their employees just to stay in business. About three million Americans soon found themselves without jobs or other income. The Great Depression was under way.

People flooded onto New York City's Wall Street, the center of the country's financial sector, after the stock market crashed.

Diamond National Bank in Pittsburgh, Pennsylvania

CHAPTER 3

BANKS GO BUST

People thought their money was safe in banks, but the Great Depression proved otherwise.

MILLIONS OF AMERICANS KEPT THEIR savings in banks, where they believed the money would be safe. Desperate for cash as the Depression took hold, a huge number of people rushed to their banks at the same time to take out their savings. However, banks kept a portion of their **capital** in investments and loans, and they didn't have enough cash on hand to pay so many customers at once. Many of them closed temporarily, and some even went out of business.

On December 11, 1930, the Bank of the United States in New York City went out of business, taking with it the savings of about 500,000 depositors. In the first three years of the Depression, nine million savings accounts were lost and several thousand banks went under. President Herbert Hoover created the Reconstruction Finance Corporation to assist and save banks, railroad companies, and state governments that were in financial trouble. Although Hoover's actions were well-meaning, they were too little too late.

People tried to get into banks to withdraw their savings, but many banks closed.

SAFE DEPOSIT
VAULTS
AMERICAN
UNION
BANK
4 % PAID ON
THRIFT ACCOUNTS
AMERICAN
UNION
BANK

"I am convinced we have now passed the worst, and with continued unity of effort, we shall rapidly recover."

—President Herbert Hoover, 1930

CHAPTER 4

A PRESIDENT FALTERS

The country was unprepared for the Depression, and so was its president.

HERBERT HOOVER BEGAN HIS PRESIDENCY IN March 1929 as one of the most admired men in America. But then came the stock market crash just seven months after his **inauguration**. Hoover declared that the difficulties would quickly pass despite "a number of persons thrown temporarily out of work." However, when the economy didn't improve, he finally began to take action. In addition to creating the Reconstruction Finance Corporation, Hoover cut taxes to stimulate spending and initiated public works projects that provided people with jobs. Among these projects was the completion in 1931 of the Empire State Building in New York, the world's tallest building at the time.

Though Hoover's actions helped many people, more radical measures were needed to truly solve the country's economic problems. Hoover, however, believed that giving direct government aid to people and companies was wrong. As a result, federal aid was withheld and the Depression deepened. As the economy got worse and worse, Hoover's popularity also plummeted.

The construction of the Empire State Building employed 3,400 workers at any given time.

Herbert Hoover waves to crowds in Elizabeth, New Jersey, during his 1928 campaign for president.

ELIZABETH
ELECTRICAL
FIXTURES
SHOES
TOWNLEY'S
RITZ
CLOTHES
FOX
2

With few jobs available, unemployed people did everything they could to gain the attention of potential employers.

CHAPTER 5

EKING OUT A LIVING

Out of work and out of money, people were willing to do almost anything to survive.

BY 1933, ABOUT 13 MILLION AMERICANS WERE out of work. Men who had once held high-paying jobs were suddenly standing in breadlines to get free meals from private charities or the government. Desperate for work so they could provide for their families, men put aside their pride and took any jobs they could find. One company executive went to work as a golf caddy at the country club where he was once a member. Other men sold cheap trinkets and toys, setting up shop in the display windows of closed stores. Some took advantage of the surplus of apples grown in the Northwest and bought crates of the fruit to sell on street corners for a nickel apiece. Their slogan was "Buy an apple a day and eat the depression away!"

Many men offered nothing but themselves. They walked the streets with handmade signs hanging from their necks. The signs described their job experience and pleaded for work. "WANTED: A Decent JOB by a Decent Man," read one heartbreaking sign.

Breadlines providing free food to the poor and hungry became a common sight across America. At one point, there were 82 breadlines in New York City alone.

Breadlines in major cities could stretch for several blocks.

Veterans made their way to Washington, D.C., from as far away as California.

CHAPTER 6

THE BONUS ARMY ON THE MARCH

The Bonus Army began in hope and ended in disappointment and chaos.

SOME PEOPLE TURNED TO THE GOVERNMENT for assistance. These included thousands of **veterans** of World War I (1914–1918) who wanted the bonus money the government had promised them after the war. The bonuses were meant to make up for the money soldiers were unable to earn at their jobs while serving in the military. The vets had been given certificates that would be paid in 1945. But they felt that in desperate times, they deserved to be paid immediately.

A thousand veterans marched to Washington, D.C., in May 1932. Within a month this "Bonus Army" was joined by 20,000 more vets and their families. They set up a **shantytown** near the Capitol while awaiting Congress's decision. In late June, Congress voted against early payment of the bonus. Discouraged, most of the veterans went home. But by late July, there were still about 2,000 people remaining. President Hoover ordered the army to drive them out. The vets and their families fled in panic as soldiers burned their shanties. The chaos led to **casualties**, including an infant who died from inhaling tear gas.

Veterans wait on the front lawn of the U.S. Capitol for Congress to debate the issue of early payment of their bonuses.

This family in Amarillo, Texas, lived in a homemade shack with no running water or bathroom.

CHAPTER 7

THE SHAME OF A NATION

Evicted from their apartments and houses, Americans created new homes and communities from wood, elbow grease, and ingenuity.

THE BONUS ARMY'S SHANTYTOWN IN Washington was just one of hundreds of similar settlements that arose in cities and towns across America during the Depression. Families that had been evicted from their homes because they couldn't make rent or **mortgage** payments shared space in these makeshift communities. People called the shantytowns Hoovervilles as a dig at the president they blamed for their suffering. Hoovervilles often sprang up near railroad yards. People tried to make their dwellings as much like home as they could by bringing furniture from their old houses and apartments.

Some homeless people who couldn't build shanties of their own moved into abandoned trolley cars, huge unused pipes, and even abandoned ovens at closed steel plants. The Hoovervilles that survived the Depression were torn down by the early 1940s.

"Hoover blankets" were the old newspapers homeless people slept under. "Hoover hogs" were rabbits that hungry people shot for food, and "Hoover flags" were the empty pockets of the poor pulled inside out.

Thousands of people lived in this Hooverville along the waterfront in Seattle, Washington.

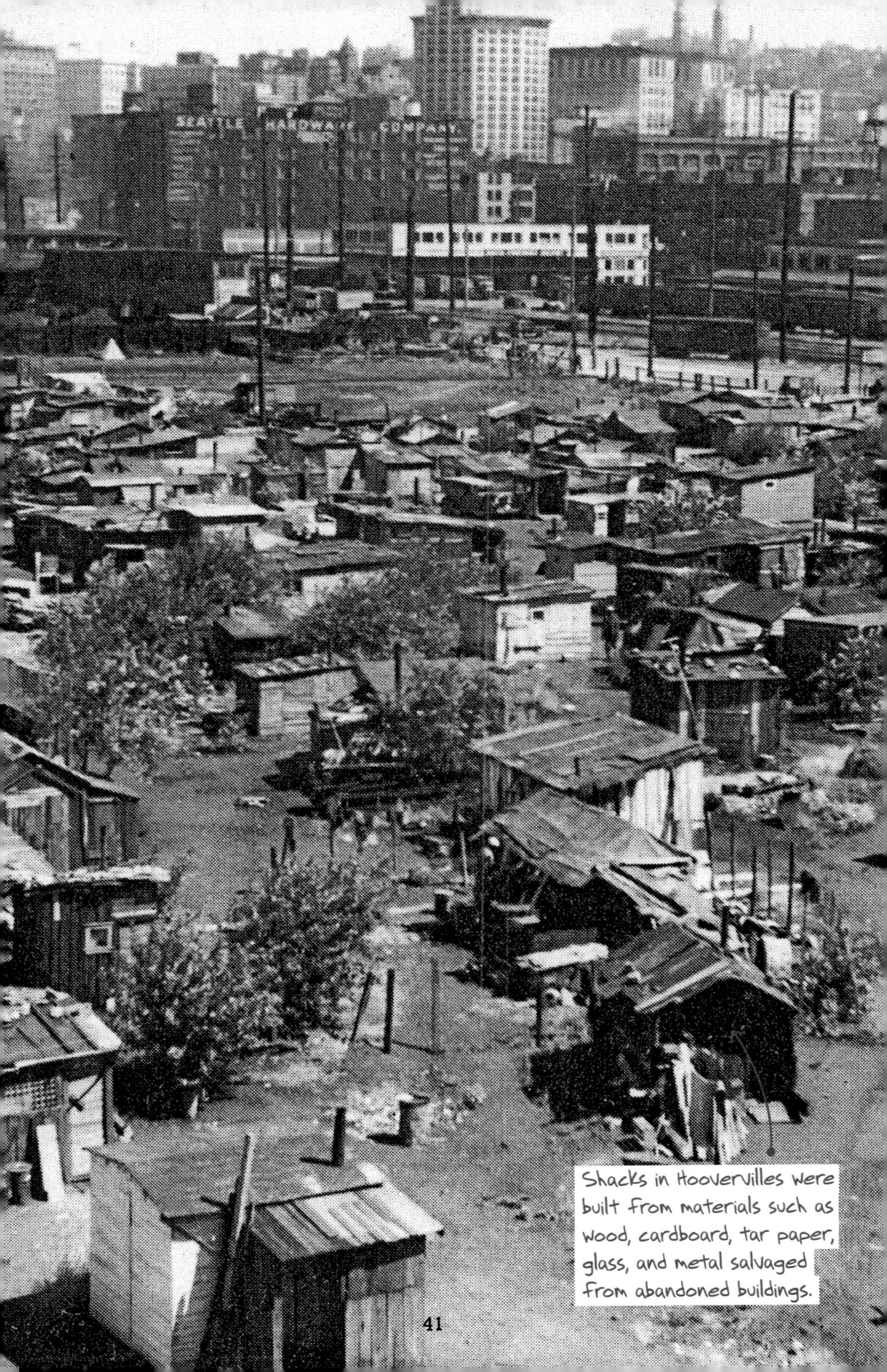

Shacks in Hoovervilles were built from materials such as wood, cardboard, tar paper, glass, and metal salvaged from abandoned buildings.

Some families had many children to feed, but little money for food.

CHAPTER 8

CHILDREN OF THE DEPRESSION

A carefree childhood was denied to millions of kids during the Depression.

YOUNG PEOPLE WERE THE MOST VULNERABLE victims of the Great Depression. Unemployed parents couldn't afford food or decent clothing for their children, who went hungry and lacked warm clothes in the winter. In some regions of the country, such as West Virginia and Pennsylvania, as many as 90 percent of all children suffered from malnutrition. Some parents couldn't afford to keep their children at home and sent them off to live with relatives, breaking up families.

But even children whose families were better off and remained intact faced challenges. Many quit school to go to work and help support their families. Others found themselves unable to attend school even if they wanted to. Thousands of schools across the country were forced to close because of a lack of money. As many as 40 percent of all youths from ages 16 to 24 were neither in school nor working. Some got into trouble and were involved in petty crime or worse. Depression-era children, in too many cases, missed out on childhood altogether.

A woman and her children pose for famed photographer Dorothea Lange at a shantytown in Oklahoma.

Hoboes often ate simple foods that could be cooked over campfires.

CHAPTER 9

RIDING THE RAILS: HOBO LIFE

Hoboes made a living by working odd jobs, but living on the open road was a constant struggle.

DURING THE GREAT DEPRESSION, MANY unemployed young people took up the hobo life. Hoboes, often called "gentlemen of the road," were migrant workers who looked for work wherever they could find it. In exchange for doing odd jobs for people, they got a little money, a place to sleep, and meals.

At the height of the Depression, it is estimated there were more than two million male hoboes and as many as 8,000 female hoboes. Without cars to get around the country, they "rode the rails," sneaking into boxcars on trains for a free ride to the next job. It was a life of freedom and adventure, but also much danger. Many people lost limbs or even died jumping onto or off of moving trains. Then there were the "bulls," vicious guards hired by the railroad companies to keep hoboes from stealing rides.

Between train rides, hoboes congregated in urban shantytowns called "hobo jungles" or in open encampments in the countryside. They'd pitch tents and gather around a campfire at night to play music and exchange stories about their adventures on the road.

Hoboes carried all of their possessions with them wherever they went.

ERIE

Listening to a favorite record was one way people could briefly forget about the troubles of the Depression.

CHAPTER 10

SONGS OF HOPE AND HAPPINESS

Music had the power to lift people's spirits, even in the darkest of times.

POPULAR MUSIC REFLECTS ITS TIME, AND this was certainly true during the Great Depression. For example, folk singer-songwriter Woody Guthrie performed his songs about the troubles and joys of common people in hobo encampments. Among his best-known songs are "Hard Traveling" and "This Land Is Your Land."

Pop songwriters also left their mark. Such upbeat tunes as "Life Is Just a Bowl of Cherries," "We're in the Money," and "On the Sunny Side of the Street" put smiles on listeners' faces when life was anything but easy.

The song that defined the era was a more grimly realistic number about a working man reduced to begging. "Brother, Can You Spare a Dime?" was written by lyricist E. Y. "Yip" Harburg and composer Jay Gorney. Gorney claimed the tune was based on a Russian-Jewish lullaby his mother sang to him as a child. **Conservatives** hated the song because it seemed to show the failure of **capitalism** during the Depression. Many tried unsuccessfully to get it banned from the radio.

Harburg wrote the lyrics for "Over the Rainbow" and the other memorable songs in the classic 1939 film *The Wizard of Oz*.

Woody Guthrie performs from a New York City stoop for a group of passersby in 1943.

KILLS

Democrat Franklin Delano Roosevelt (left) faced off against incumbent Republican Herbert Hoover (right) in the 1932 presidential election.

CHAPTER 11

THE ELECTION OF 1932

The country was ready for a change in government as the election of 1932 approached.

IN 1932, IT WAS CLEAR THAT REPUBLICAN president Herbert Hoover could be easily beaten in his bid for reelection. A number of Democrats vied for the presidential nomination. Two-term New York governor **Franklin Delano Roosevelt** emerged as the winner.

As governor, Roosevelt had been resourceful in dealing with the Depression in his state. However, he was also paralyzed from the waist down and confined to a wheelchair. Some people felt he didn't have the stamina or good health to be president. But FDR—as he was popularly known—surprised his critics with his tireless campaigning and dynamic speeches in which he promised, unlike Hoover, to provide federal assistance and reform programs for all Americans in need.

In his acceptance speech for the nomination, Roosevelt said, "Give me your help, not to win votes alone, but to win in this crusade to restore America to its own people." He was elected in a landslide. The Democrats also easily won both houses of Congress. Roosevelt had promised to solve the nation's problems. Now the nation waited to see if he would make good on this promise.

Roosevelt's paralysis was a result of his battle with the disease called polio in 1921, when he was 39 years old.

President Roosevelt sometimes rode in an open-top car while campaigning to prevent people from noticing that his legs were paralyzed.

“*The only thing we have to fear is fear itself.*”

—President Franklin D. Roosevelt in his first inaugural speech on March 4, 1933

CHAPTER 12

NEW PRESIDENT, NEW DEAL

The first hundred days of Roosevelt's presidency were unlike any seen before or since.

ROOSEVELT WASTED NO TIME IN GOING TO work on a "new deal for the American people." This New Deal, as it came to be called, was a set of federal programs that aimed to put Americans back to work and start up the economy. On his first day in office, Roosevelt declared a four-day "bank holiday." All banks shut down while Congress worked out the Emergency Banking Act, which helped them reopen on a solid foundation. It was just one of 15 major laws that Congress passed during the first 100 days of Roosevelt's presidency.

Some of the New Deal programs and agencies were very successful. Among them was the Civilian Conservation Corps, which put more than two million unemployed people to work on conservation projects, including building roads and reforesting land. The National Recovery Administration (NRA) worked to create fair business practices that would stimulate the economy. The Works Progress Administration (WPA) put people to work building schools, hospitals, and bridges. From 1935 to 1943, when it disbanded, the WPA gave work to 8.5 million Americans. Roosevelt's New Deal was bringing new hope to the country.

WPA workers build a tunnel in Mobile, Alabama.

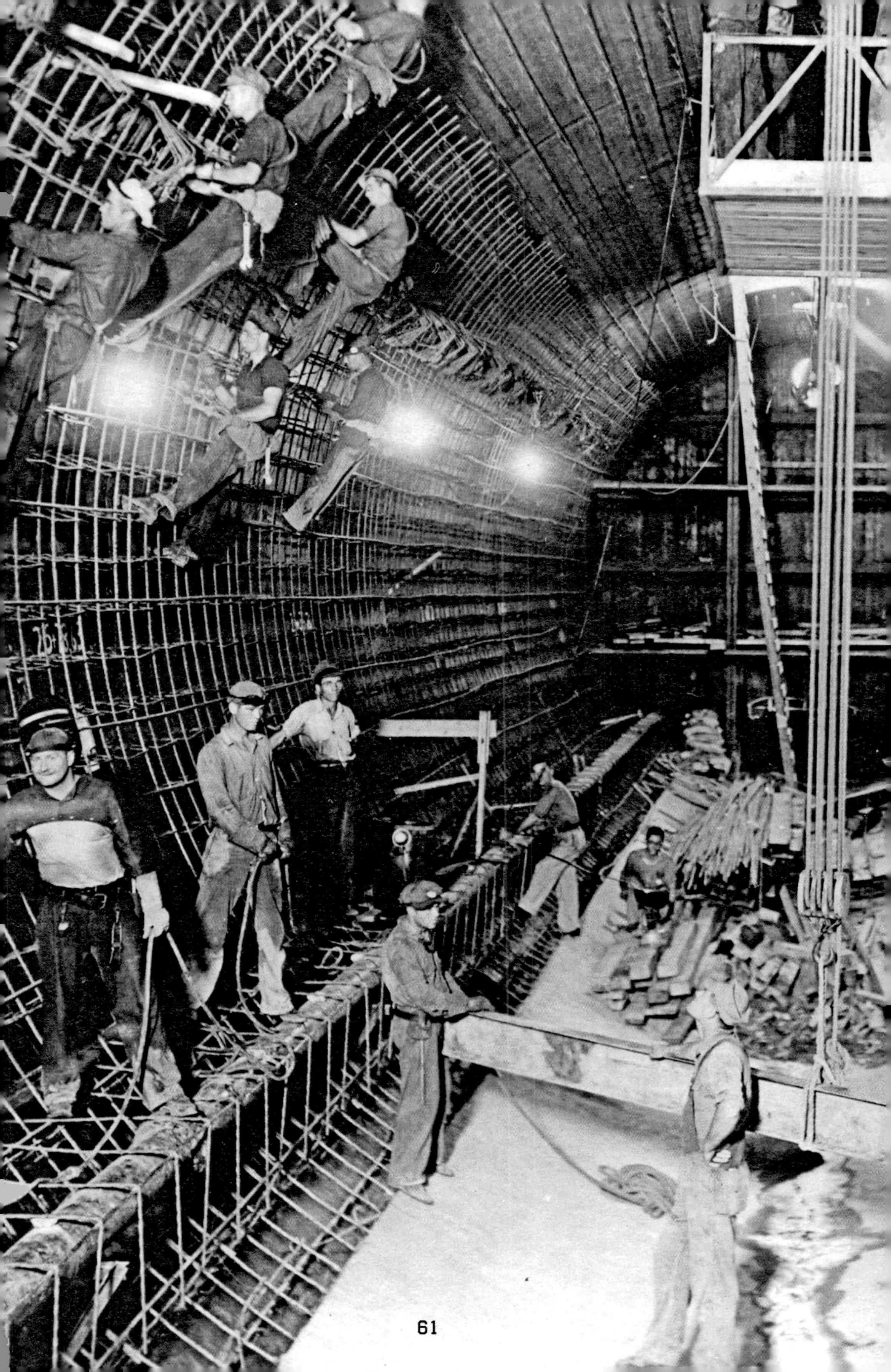

Workers install a massive electrical generator in Tennessee as part of a New Deal project.

CHAPTER 13

DAMS, ROADS, AND BRIDGES

Thousands of workers helped build a series of architectural wonders during the Great Depression.

IMPROVING THE NATION'S **INFRASTRUCTURE** with bridges, buildings, roads, and dams was a major goal of the New Deal.

Among New Deal projects, one of the most successful was the Tennessee Valley Authority (TVA), established in 1933. The Tennessee River basin stretched across seven states and was plagued by annual flooding. The TVA put 15,000 people to work clearing forestland and building dams and roadways. This work helped control flooding, improved navigation along the river, and provided cheap electricity generated from hydropower.

Another architectural wonder of the Depression was the Hoover Dam on the Colorado River, along the border between Nevada and Arizona. Begun under President Hoover in 1931, the project employed thousands of workers at a time. The dam was completed in 1936. Still standing today, it remains one of the highest concrete dams in the world. Its construction stopped flooding along the Colorado and brought water to the Southwest, turning a desert into a fertile region.

A worker sits in a truck being lifted above the construction site of the Hoover Dam.

President Roosevelt broadcasts a speech to all the major radio networks from his desk in the White House in 1944.

CHAPTER 14

FIRESIDE CHATS AND BELLY LAUGHS

Roosevelt wasn't the first president to speak on the radio, but he was the first to master the practice as a means of mass communication.

BY THE END OF THE 1930S, 85 PERCENT OF all American households owned a radio. Just days after his inauguration, President Roosevelt went on the radio to talk to the American people about his new banking program. It was the first of 30 "fireside chats" by the president. Roosevelt typically started his chats with the words "My friends." Listeners in their homes felt like the president was talking directly to them. He was no longer just their president, but a personal friend who cared about their welfare.

Radio delivered far more than news and speeches. It was an endless stream of free entertainment that took people's minds off troubles such as a lack of money and jobs and an uncertain future. In the morning, homemakers doing their chores tuned in to hear their favorite daily dramas, or soap operas. After school, children rushed home to listen to the latest adventures of Little Orphan Annie or the detective Dick Tracy. After dinner, the whole family gathered around the radio to hear the day's top comedy and variety shows.

They were called soap operas because many of the shows' sponsors were soap companies.

Children gather to listen to a special speech from President Roosevelt in which he addressed the nation's youth.

PHILCO

DECEMBER 5, 1933

Prohibition is repealed.

AUG SEPT OCT NOV DEC **1934** JAN FEB

CHAPTER 15

HAPPY DAYS ARE HERE AGAIN

For the first and only time in U.S. history, an **amendment** to the U.S. Constitution was repealed, and most Americans agreed that it was a good idea.

WHILE CONGRESS WAS BUSY MAKING new laws to fight the Depression, they also took time to undo one law. **Prohibition** banned the sale of alcoholic beverages. It had passed as the 18th Amendment to the Constitution in 1919 and went into effect the following year.

By all accounts, Prohibition had been a failure. Gangsters supplied illegal booze to whoever wanted to keep drinking. But this alcohol was cheaply made and contained ingredients that were often harmful. Bad liquor was the cause of an estimated 1,000 deaths a year and many other serious illnesses.

On December 5, 1933, three-fourths of the states **ratified** the 21st Amendment—to repeal Prohibition—making it law. Americans celebrated by having a drink at their favorite bar or restaurant. President Roosevelt, who favored the repeal, expressed the feelings of millions of Depression-weary Americans when he declared, "What America needs now is a drink."

On October 28, 1932, more than 20,000 people gathered in Newark, New Jersey, to protest Prohibition.

WE
WANT
BEER
1933
WE
WANT
BEER
1933
WE
WANT
BEER
WE
WANT
BEER
1933

Some people wore special masks to help them breathe during dust storms on the Great Plains.

CHAPTER 16

LOST IN THE DUST BOWL

Nature itself seemed determined to bring the farmers of the Great Plains to ruin.

Go to page 13 to see a map of the Dust Bowl.

FEW PEOPLE WERE HIT AS HARD BY THE Depression as farmers. Crop prices plummeted because people couldn't afford to buy food. This drove thousands of small farmers into bankruptcy, causing them to lose their farms. To make matters worse, the Great Plains were affected by severe droughts in the early 1930s, causing crops to wither. Then, in November 1933, a terrible dust storm swept across South Dakota. The air was so thick with dust that people called it a black blizzard. It was the first of a series of devastating dust storms that would persist for two years. During these dust storms, millions of acres of farmland were ruined by erosion and lack of rain. They formed an enormous Dust Bowl.

Thousands of families left their homes and farms, strapped their belongings onto automobiles, and headed for California. As many as 350,000 people were part of this great migration. They hoped to find new opportunities in the fertile valleys of California. But most found only backbreaking fieldwork for meager pay.

Native Californians looked down on the newcomers and called them Okies, although many came from states other than Oklahoma.

A father and son walk on top of the thick layer of dust that has covered their farm.

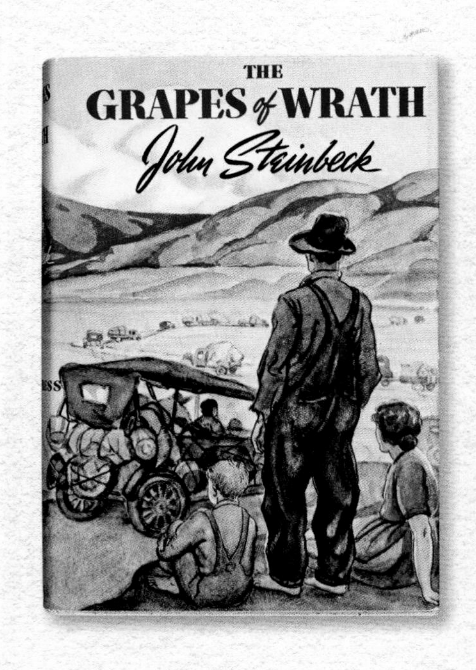

John Steinbeck's The Grapes of Wrath *is a classic tale of the Depression's effect on farmers in the Dust Bowl.*

CHAPTER 17

DEPRESSION WRITERS

Some of America's greatest writers made their mark writing about this tumultuous time.

THE PLIGHT OF THOSE WHO FLED THE DUST Bowl was memorably captured in John Steinbeck's best-selling novel *The Grapes of Wrath*. The book didn't appear until 1939, when the Depression was nearing its end, but no other work of literature so vividly captured the determined spirit of the Dust Bowl migrants. The novel won the Pulitzer Prize for Fiction and was made into a classic film by director John Ford the following year. The movie starred Henry Fonda as Tom Joad, the son of the migrant family Steinbeck's story focuses on.

Other American writers expressed the hopes and dreams of Depression-era America. Playwright Clifford Odets wrote about a labor strike in his play "Waiting for Lefty." John Dos Passos chronicled the decades leading up to the Depression using a unique blend of fiction, biography, and newspaper clippings in a trilogy of novels he titled *U.S.A.* Other lesser-known writers were commissioned by the WPA to celebrate America by writing colorful state guidebooks that promoted tourism.

In 1962, Steinbeck was awarded the Nobel Prize in Literature, one of the highest honors a writer can receive.

Henry Fonda (center) was nominated for an Academy Award for his portrayal of Tom Joad in The Grapes of Wrath.

Even when it came to things as simple as soda machines, black people faced unfair treatment.

CHAPTER 18

THE HARDEST HIT: AFRICAN AMERICANS

Poverty and racism only grew worse for black Americans during the Depression.

OF ALL MINORITY GROUPS, AFRICAN Americans suffered the most during the Great Depression. In the South, plunging cotton prices put black farmers out of business and left **sharecroppers** without work. In both northern and southern cities, black people lost the jobs they had traditionally held—busboys, elevator operators, porters, and maids—so the jobs could be given to unemployed white workers. Urban African Americans had an unemployment rate of more than 50 percent, twice that of urban whites. Those who could find work, especially women, were paid very little.

President Roosevelt's New Deal was seen as a raw deal for African Americans. Roosevelt did little to help blacks, afraid that if he did, he would lose the support of **segregationist** southern senators. His wife, Eleanor, in contrast, was a social activist who reached out to black communities. Her influence eventually caused her husband to do more for African Americans in his second term. He even formed a "black cabinet" of advisers that included the black educator Mary McLeod Bethune.

Because of racist attitudes, some charities would not serve black people at their free soup kitchens.

Louisiana sharecroppers with bags of recently picked cotton

GET·DILLINGER!

$15,000 Reward

A PROCLAMATION

WHEREAS, One John Dillinger stands charged officially with numerous felonies including murder in several states and his banditry and depredation stamp him as an outlaw, a fugitive from justice and a vicious menace to life and property;

NOW, THEREFORE, We, Paul McNutt, Governor of Indiana; George White, Governor of Ohio; F. B. Olson, Governor of Minnesota; William A. Comstock, Governor of Michigan; and Henry Horner, Governor of Illinois, do hereby proclaim and offer a reward of Five Thousand Dollars ($5,000.00) to be paid to the person or persons who apprehend and deliver the said John Dillinger into the custody of any sheriff of any of the above-mentioned states or his duly authorized agent.

THIS IS IN ADDITION TO THE $10,000.00 OFFERED BY THE FEDERAL GOVERNMENT FOR THE ARREST OF JOHN DILLINGER.

HERE IS HIS FINGERPRINT CLASSIFICATION and DESCRIPTION. —— FILE THIS FOR IDENTIFICATION PURPOSES.

John Dillinger, (w) age 30 yrs., 5-8½, 160½ lbs., gray eyes, med. chest, hair, med. comp., med. build. Dayton, O., P. D. No. 10587. O. S. B. No. 559-646.

F.P.C. (12) M 9 R O O

FRONT VIEW

SIDE VIEW

Be on the lookout for this desperado. He is heavily armed and usually is protected with bullet-proof vest. Take no unnecessary chances in getting this man. He is thoroughly prepared to shoot his way out of any situation.

GET HIM
DEAD
OR ALIVE

Notify any Sheriff or Chief of Police of Indiana, Ohio, Minnesota, Michigan, Illinois.

or THIS BUREAU

ILLINOIS STATE BUREAU OF CRIMINAL IDENTIFICATION AND INVESTIGATION
T. P. Sullivan, Supt.
Springfield, Illinois

Law enforcement officials offered huge rewards to anyone who could help capture dangerous bank robbers such as John Dillinger.

CHAPTER 19

PUBLIC ENEMY NUMBER ONE

To law enforcement, bank robbers were public enemies, but to many people, they were heroes.

THE REPEAL OF PROHIBITION MEANT THE end of the bootlegged liquor industry. Many criminals turned to robbing banks to make their living. Some bank robbers became heroes to many ordinary Americans, who had come to hate banks. The most famous of them was John Dillinger of Indiana. In a series of daring robberies and jailbreaks, Dillinger rose to become number one on the Federal Bureau of Investigation's (FBI) list of most wanted criminals.

Perhaps the most admired gangster was **Charles Arthur "Pretty Boy" Floyd**. When robbing a bank with his gang, Floyd would find the bank's mortgages and tear them up so the banks could not take people's homes. As he sped away, he scattered dollar bills out the window of his getaway car for people to find.

Crime paid big for these gangsters, but only for a time. Dillinger was shot down by FBI agents leaving a Chicago movie theater. Floyd was shot down in an Ohio cornfield while fleeing a 100-man posse.

Floyd and Dillinger died within three months of each other in 1934.

The notorious criminals Bonnie and Clyde often took photos of themselves as they traveled around the country robbing banks and murdering people.

587-

“*Photography takes an instant out of time, altering life by holding it still.*”

—Dorothea Lange

CHAPTER 20

THE DEPRESSION IN PICTURES

The saying “A picture is worth a thousand words” was dramatically proven in the expressive work of Depression-era photographers.

NOTHING EXPRESSED THE SUFFERING and unconquered spirits of the neediest Depression-era Americans in a more compelling way than photographs. Dorothea Lange was a young portrait photographer in San Francisco who was hired to photograph migrant workers and others by the Farm Security Administration (FSA). One of her photographs, *Migrant Mother*, is the most famous image of the era. According to Lange, the mother "sat in that lean-to tent with her children huddled around her, and seemed to know that my pictures might help her, and so she helped me."

Another celebrated FSA photographer was Walker Evans, whose photographs of southern sharecroppers brought their plight to national attention. He and writer James Agee spent time with three Alabama sharecropper families and later collaborated on a book about their experiences, *Let Us Now Praise Famous Men*. It quickly became a Depression classic. Both Lange and Evans went on to have long and productive careers after the Depression ended.

Dorothea Lange's photo of a California mother and her children is one of the most famous images of the Great Depression.

German children wave flags supporting the Nazi Party in 1934.

CHAPTER 21

THE RISE OF DICTATORS

The worst of times brought out the worst of leaders in several European nations.

THE DEPRESSION DIDN'T STOP AT THE borders of the United States. It spread over the entire world. Europe was hit especially hard. Germany, still recovering from its humiliating defeat in World War I, saw massive unemployment and soaring inflation. The German people turned in desperation to Nazi Party leader **Adolf Hitler**, who promised to return Germany to its former greatness. Hitler, a ruthless dictator, started rearming Germany for war, which brought jobs and money to the German people, but also led to World War II (1939–1945). Italy also experienced economic difficulties, and a decade earlier had turned to a dictator, Benito Mussolini, to save the country.

Meanwhile in Spain, a civil war erupted in 1936. The ruling Republicans were pitted against the Nationalist rebels, led by **General Francisco Franco**. The Nationalists won the war in 1939, and Franco joined Hitler and Mussolini as one of Europe's most powerful dictators. Pressured by the United States, Franco did not join Germany and Italy in World War II. He managed to stay in power for the next 35 years, until his death in 1975.

Italy's Benito Mussolini (left) and Germany's Adolf Hitler (right) were brutal dictators whose leadership led to the deaths of millions of people.

“I’m for the poor man—all poor men, black and white, they all gotta have a chance.”

—Huey Long

CHAPTER 22

THE KINGFISH

He was a powerful force in American politics whose ideas for economic recovery went beyond the New Deal.

THE GREAT DEPRESSION PRODUCED NO dictators in America, but it did nurture some **demagogues**. The most prominent of them was **Huey Long** of Louisiana. Long had little formal education but was a smooth talker. He had been a successful traveling salesman before turning to politics. In 1928, he was elected the youngest governor in Louisiana history. He ran the state with an iron hand. He aligned himself with the poor and did much good, such as building a highway system, funding colleges, and providing free textbooks for schoolchildren.

Nicknaming himself the Kingfish after a character on the radio show *Amos and Andy*, Long entered the Senate in 1932. There he laid down a program to redistribute the nation's wealth by raising taxes on the wealthy. Long soon declared himself a presidential candidate, posing a real threat to Roosevelt's reelection. But on September 8, 1935, he was shot in the corridors of Louisiana's capitol by a young doctor whose family was among his many enemies. Long's bodyguards shot the assassin dead, and Long died two days later from his wounds.

Though Long was planning a presidential campaign for 1936, he had supported President Roosevelt's election in 1932.

LIBERTY
IN GOD WE TRUST
1936

UNITED STATES OF AMERICA
E PLURIBUS UNUM
ONE DIME

A 1936 dime

CHAPTER 23

THE SOCIAL SECURITY ACT OF 1935

A new program improved financial security for retired Americans.

MILLIONS OF RETIRED AMERICANS LOST their savings in the stock market crash and had no means of support. They had only family to turn to for help. The 1935 Social Security Act was passed to support these people. It guaranteed a monthly income for retirees, based on their average earnings during their working years.

Many conservatives saw the plan as extreme **socialism** that would bankrupt the U.S. Treasury. Some businesspeople tried to scare their employees into opposing Social Security by pointing out that part of their wages would go to paying for it. "You're sentenced to a weekly pay reduction for all your working life," they wrote on printed slips they put into paycheck envelopes. "You'll have to serve the sentence unless you change it November 3 [Election Day]."

Social Security has undergone many changes over the years, but it remains a very popular program. Countless Americans have relied on it to help them stay financially secure even after they have stopped working and earning money.

After the passage of the Social Security Act in 1935, the government encouraged Americans to register for the new program.

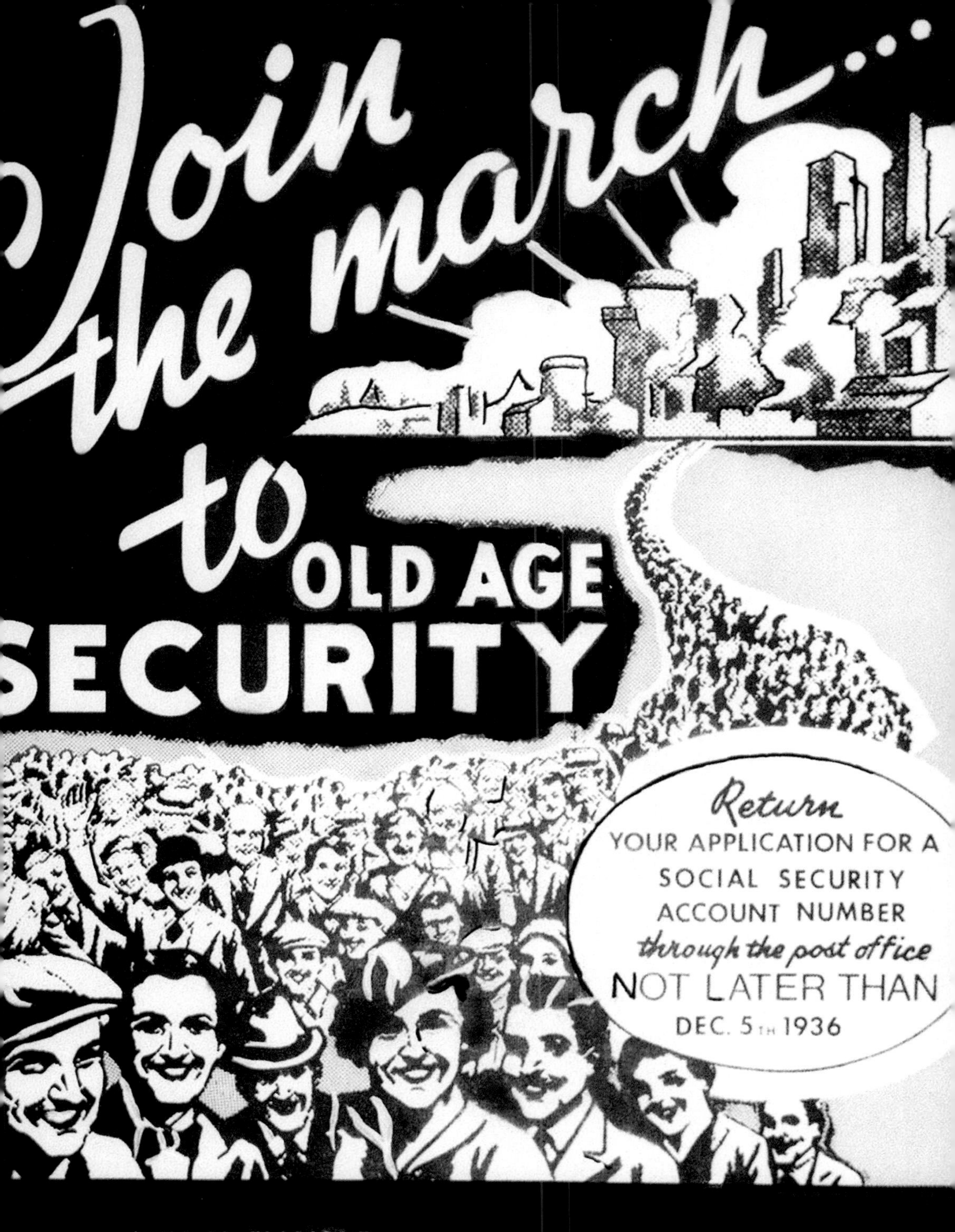

WHO IS ELIGIBLE . . . EVERYBODY WORKING FOR SALARY OR WAGES (WITH ONLY A FEW EXCEPTIONS, SUCH AS AGRICULTURE, DOMESTIC SERVICE, AND GOVERNMENT WORK). APPLICATIONS FOR SOCIAL SECURITY ACCOUNTS ARE AVAILABLE THROUGH EMPLOYERS. IF YOU DO NOT GET ONE FROM YOUR EMPLOYER, ASK FOR ONE AT THE POST OFFICE.

HOW TO RETURN APPLICATION

1. HAND IT BACK TO YOUR EMPLOYER, or
2. HAND IT TO ANY LABOR ORGANIZATION

4. DELIVER IT TO LOCAL POST OFFICE, or
5. MAIL IT IN A SEAL ENVELOPE

"There is only one issue in the campaign. It's myself, and the people must be either for me or against me."

—President Franklin D. Roosevelt,
running for reelection

CHAPTER 24

THE ELECTION OF 1936

Both the most beloved and most hated man in America, Roosevelt faced the upcoming election with confidence.

DURING HIS FIRST FOUR YEARS IN OFFICE, President Roosevelt won the support of farmers, laborers, and the working poor, all of whom benefited from the New Deal. But he also made enemies. Republicans and many wealthy Americans saw Roosevelt as a traitor to his class, a man who abandoned capitalism for socialism and liberal government giveaway programs. Eager to put "that man in the White House" out of office, Republicans nominated Kansas governor **Alf Landon** for president. A moderate Republican, Landon lacked Roosevelt's charisma and energy.

The Republican campaign slogan was "Landon by a Landslide," but on Election Day it was Roosevelt who won the landslide, taking every state but Maine and Vermont. The economy was not prospering, but millions of Americans were doing better than they had been, thanks to Roosevelt's reforms and programs. Americans had confirmed their belief in Roosevelt's policies and returned him to the presidency to finish the work of ending the Depression. As historian Paul Boller put it, "In 1932 people voted against Hoover. In 1936 they voted for FDR."

Roosevelt supporters in Philadelphia, Pennsylvania, carry posters bearing the president's face during a 1936 campaign event.

MISSOURI

The Marx Brothers' classic 1933 film Duck Soup *is a comedy about two countries that go to war because their leaders are trying to marry the same woman.*

CHAPTER 25

THE HOLLYWOOD DREAM MACHINE

Depression-era America wanted escapist entertainment, and Hollywood was happy to provide it.

THE MOVIES GREW RAPIDLY IN POPULARITY during the Depression years. They offered a feast for the eyes and ears that attracted 85 million people a week. For a few cents, moviegoers got to see a double feature, a newsreel, a cartoon, a short film, and previews of coming attractions. They might also get a chance to win a set of dishes or another door prize in a raffle.

The films themselves ranged from musicals and comedies to romantic dramas. Glitzy musicals with dozens of chorus girls directed by Busby Berkeley vied with films starring the dancing team of Fred Astaire and Ginger Rogers.

Comedies featured the madcap Marx Brothers and the more down-to-earth comedy duo of Laurel and Hardy. Young moviegoers identified with teenage actor Mickey Rooney, who starred as Andy Hardy in a series of popular films. Not all Depression-era movies were escapist. There were also serious dramas that dealt with the issues of the day, such as an unjust penal system (*I Was a Fugitive From a Chain Gang*) and homeless youth (*Wild Boys of the Road*).

Fred Astaire and Ginger Rogers dance in a scene from the 1935 film Top Hat.

NOVEMBER 1938

The Congress of Industrial Organizations (CIO) is founded by John L. Lewis and others

JULY AUG SEPT OCT NOV DEC **1939** JAN

CHAPTER 26

THE POWER OF ORGANIZED LABOR

Organized labor unions empowered American workers and gave them an advantage in dealing with their employers.

LABOR UNIONS HELPED AND PROTECTED American workers by negotiating with employers for better wages, benefits, and working conditions. But the Depression robbed unions of much of their power. Unemployment reduced the ranks of workers who were able to join unions. Those workers still employed were willing to put up with poor treatment just to keep their jobs.

Union leader **John L. Lewis** refused to let workers be taken advantage of in those dark days. He helped organize the first sit-down strike in the United States in late 1936. Workers at a General Motors factory in Flint, Michigan, went on strike to, among other things, establish a fair minimum wage and set procedures to protect workers on the assembly line from injury. When the strike began, the workers simply sat down at their work stations and refused to work. As the strike spread, General Motors finally agreed to recognize and work with the union. The following year, Lewis and other labor leaders established the Congress of Industrial Organizations (CIO), which continued to represent skilled and unskilled workers until 1955.

Workers mark the days of their sit-down strike on the side of a car at the General Motors plant in Flint, Michigan.

SIT DOWN STRIKE
SUN MON TUES WED THUR FRI SAT
10 11 12 13
14 15 16 18 19 20
21 22 23 26 27
28

Artists Phillip Reisman (left) and John Lamont (right) work on a mural commissioned by the WPA in New York City's American Museum of Natural History.

CHAPTER 27

MURALS ACROSS AMERICA

These huge paintings were among the most public of artworks. They told the story of America to all who saw them.

THE WPA'S FEDERAL ART PROJECT, ONE OF the most fondly remembered programs of the New Deal, employed musicians, theater professionals, and writers. But some of the WPA's most impressive work came from fine artists. Over a decade, 5,000 artists produced 225,000 artworks. The most enduring of these works were **murals**. Artists were commissioned by the Federal Art Project to create murals in post offices and other public buildings across the country. These huge murals celebrated events from American history and folklore and scenes of everyday life and work in local communities, often including dozens of characters.

Many of the mural artists were unknown at the time, but some were already established painters, such as Ben Shahn and Thomas Hart Benton. Others, such as Jackson Pollock, Willem de Kooning, and Mark Rothko, went on to become famous after leaving the WPA. Many of the hundreds of murals were destroyed, but a number still remain and can be seen today.

WPA artist Victor Arnautoff created this mural showing people on the streets of San Francisco, California. It is displayed on the wall's of San Francisco's Coit Tower.

CITY LIGHTS
SODA
DRUGS FOUNTAIN CANDIES
KODAKS
FILMS
LUNCH
TAVERN
The Argonaut
NEW YORK STOCKS
TIME

A poster advertising the 1939 world's fair

CHAPTER 28

THE WORLD OF TOMORROW

Weary of the past and uncertain of the present, Americans looked to the future at the 1939 world's fair in New York City.

IN 1939, NEW YORK CITY HELD A WORLD'S FAIR to celebrate the 150th anniversary of George Washington's inauguration as president. The fair's theme was not the past but "the World of Tomorrow." The fair's 45 million visitors could walk through a hydroelectric dam, meet a talking 7-foot-tall robot called Elektro, and view the city of the future in 1960 from moving chair-cars in General Motors' Futurama exhibition. They left the exhibit with buttons that read "I Have Seen the Future."

No new invention was more popular with fairgoers than television. The National Broadcasting Company (NBC), a radio broadcaster, was the pioneer of this new technology. On April 30, the fair's opening day, NBC broadcast President Roosevelt's opening speech on special screens. People several miles away at Radio City Music Hall also watched the president via NBC's transmitter atop the Empire State Building. The fair was just the boost Americans still struggling with the Depression desperately needed.

By the end of the year, about 2,500 people in Greater New York owned television sets, but only a few hours of programming aired each day.

A musician plays the cello as a robot appears to conduct the music behind her at the 1939 world's fair.

The Japanese military used fighter planes like this one during World War II.

CHAPTER 29

A NEW WAR

A new world war brought death and destruction, but it also completed America's emergence from the Great Depression.

ON SEPTEMBER 1, 1939, GERMANY INVADED Poland. Two days later, Great Britain and France declared war. Soon all of Europe was enveloped in the war. While the United States was concerned, it had too many problems of its own to get involved. The New Deal's effects had slowed down, and unemployment was rising again.

Roosevelt decided to run for a third term in 1940. His Republican opponent, Wendell Willkie, called the president a "warmonger," but Roosevelt continued to keep the country out of the conflict in Europe. He won the election.

Then, on December 7, 1941, Japan—an ally of Germany—launched a sneak attack on the U.S. naval base at Pearl Harbor in Hawaii. At Roosevelt's urging, Congress declared war on Japan and later Germany as well. American companies went to work churning out **armaments** and supplies for American soldiers. Factories needed workers, and the unemployment rate plummeted. Within a year of the Pearl Harbor attack, the Great Depression, after 13 years, was officially declared to be over.

Women polish the nose cones of airplanes at a factory during World War II.

"No one could possibly have lived through the Great Depression without being scarred by it."

— Author Isaac Asimov

CHAPTER 30

THE LEGACY OF A DARK TIME

The Great Depression was over, but its memory would live on.

WORLD WAR II ENDED IN VICTORY FOR the United States and its allies in 1945 and brought prosperity to Americans in the postwar years. The Great Depression soon seemed like some kind of bad dream, but it left a legacy that would linger for generations to come. Young people who came of age during the Depression would never take their future material success for granted. The same resilience that their parents showed during those dark days was reflected in their own spirit. This strength of spirit helped them fight and win World War II.

Another important part of the Depression's legacy is the role the government plays in Americans' lives. The New Deal lived on in Social Security and Medicare, federal programs that have continued to help Americans ever since. The government continues to this day to be seen as a power that can help people in their lives. America would never be the same again after the Great Depression. It was the worst of times for most Americans, but it also proved that the human spirit could survive any setback.

Just as they had on the day the stock market crashed, people flooded onto New York City's Wall Street to celebrate the end of World War II in 1945.

NASSAU ST.
PARKING SPACE U.S. GOVT CARS ONLY

KEY PLAYERS

President Herbert Hoover (1929–1933) had barely begun his presidency when the stock market crashed and the country was plunged into the Great Depression. While his intentions were good, Hoover did not take the steps needed to combat the Depression.

President Franklin D. Roosevelt (1933–1945) came into office determined to use the government and his power as president to rebuild the economy and provide jobs for millions of unemployed Americans. While his New Deal had its failures as well as its successes, his forceful personality and optimism gave Americans hope.

Huey Long, governor and then U.S. senator from Louisiana, presented a daring agenda to provide for the nation's poor that went beyond Roosevelt's New Deal and threatened the president's reelection. A spellbinding demagogue, Long was murdered before he had the opportunity to put his plans into action.

Alf Landon, Republican governor of Kansas, ran against Roosevelt in the presidential election of 1936. A good man with a modest personality, he lost in a landslide to Roosevelt, who had the support of the many Americans who were helped by his New Deal.

Charles Arthur "Pretty Boy" Floyd was one of the most infamous gangsters of the early 1930s. Floyd is still a hero to many in his home state of Oklahoma because he robbed the much-hated banks and distributed some of his stolen loot to the poor.

John L. Lewis was a union leader who fought for American workers during the Depression and long after. He was instrumental in founding a large union, the Congress of Industrial Organizations, in 1938.

Adolf Hitler was a German dictator who convinced his countrymen and women that he could save them from economic depression and return Germany to greatness. He ended up plunging his nation and much of the world into another world war.

General Francisco Franco of Spain led a rebellion against his country's democratic Republican government in 1936. Franco's Nationalists, supported by Hitler, eventually won, and Franco ruled Spain as a dictator for the next 35 years.

GREAT DEPRESSION TIMELINE

OCTOBER 24
The stock market crashes.

OCTOBER 29
The market hits bottom with $15 billion lost in stocks on so-called Black Tuesday.

DECEMBER 11
The Bank of the United States closes, taking with it the savings of 500,000 depositors.

More than eight million Americans, nearly 16 percent of the labor force, are unemployed.

JUNE
Twenty thousand veterans of World War I and their families arrive in Washington, D.C., to demand Congress give them their war bonuses.

NOVEMBER 8
Democrat Franklin Delano Roosevelt defeats President Herbert Hoover in the presidential election.

1929 J F M A M J J A S O N D 1930 J F M A M J J A S O N D 1931 J F M A M J J A S O N D 1932 J F M A M J J A S

MARCH–JUNE
The president and Congress pass 15 major laws to help restart the economy and provide jobs in Roosevelt's first 100 days in office.

MARCH 6
President Roosevelt declares a four-day "bank holiday," closing all banks until Congress can take action to help them re-open on a more solid foundation.

DECEMBER 5
Prohibition is repealed with the ratification of the 21st Amendment to the Constitution.

1933 JAN FEB **MAR** **APR** **MAY** **JUN** JUL AUG SRP OCT **NOV** **DEC** 1934 J F M A M J J A S **O** N D

MARCH 12
Roosevelt delivers the first of his "fireside chats" on the radio to the American public.

NOVEMBER
The first of a series of devastating dust storms strikes the Great Plains, a region that came to be called the Dust Bowl.

1933
Unemployment reaches a peak of 25 percent in the United States.

OCTOBER 22
Bank robber "Pretty Boy" Floyd is killed by federal agents in Ohio.

AUGUST 14
The Social Security Act is signed into law by the president.

NOVEMBER 3
President Roosevelt is reelected to a second term.

1935 J F M A M J J A S O N D **1936** J F M A M J J A S O N D **1937** J F M A M J J A S O N D **1938** J F M A M J J A

JULY
The Spanish Civil War breaks out, and hundreds of American volunteers head to Spain to fight for the Republican cause against the Nationalists led by General Francisco Franco.

NOVEMBER
The Congress of Industrial Organizations (CIO), a labor union, is founded by labor leader John L. Lewis and others.

SEPTEMBER 8
Senator Huey Long of Louisiana is assassinated at the capitol in Baton Rouge.

As America experiences a wartime economic boom, the Great Depression is officially declared over.

APRIL 30
The world's fair opens in New York City.

NOVEMBER 5
Roosevelt is elected to an unprecedented third term as president.

939 J F M A M J J A S O N D 1940 J F M A M J J A S O N D 1941 J F M A M J J A S O N D 1942 J F M A M J J A S O

SEPTEMBER 1
Germany invades Poland, setting off World War II.

DECEMBER 7
The Japanese launch a sneak attack on the U.S. naval base at Pearl Harbor, Hawaii, and the next day the United States enters World War II.

GLOSSARY

- **amendment** (uh-MEND-muhnt) *noun* a formal change or revision to a law or official document
- **armaments** (AHR-muh-muhnts) *noun* weapons and other equipment used for fighting wars
- **bankruptcy** (BANGK-ruhpt-see) *noun* the situation of not having enough money to pay back what is owed; the result is the person's or business's remaining property is distributed among those owed the money
- **capital** (KAP-ih-tuhl) *noun* wealth or money
- **capitalism** (KAP-ih-tuh-liz-uhm) *noun* an economic system where goods and means of production and distribution are privately owned
- **casualties** (KAHZ-oo-uhl-teez) *noun* people who are injured or killed in an accident, a natural disaster, or a war
- **conservatives** (kuhn-SUR-vuh-tivz) *noun* people who believe politically in things staying as they are
- **demagogues** (DEM-uh-gohgz) *noun* political leaders who gain power by appealing to the prejudices of the people
- **depression** (dih-PRESH-uhn) *noun* a time during which there is a rise in unemployment, a drop in wages and prices, and little business activity
- **inauguration** (in-aw-gyuh-RAY-shuhn) *noun* a formal ceremony of installing a president or other elected official in office
- **infrastructure** (IN-fruh-struhk-chur) *noun* a system of public works such as roads and bridges

- **mortgage** (MOR-gij) *noun* a document proving the right to property, given as security for payment of a debt
- **murals** (MYOOR-uhlz) *noun* pictures painted on a wall or ceiling
- **Prohibition** (proh-hih-BISH-uhn) *noun* period from 1920 to 1933 during which alcoholic beverages were made illegal by the 18th Amendment to the U.S. Constitution
- **ratified** (RAT-uh-fyed) *verb* validated a law by giving formal consent or approval
- **segregationist** (seg-rih-GAY-shuhn-ist) *noun* a person who supports separating groups of people from other people, especially African Americans
- **sharecroppers** (SHAIR-kraph-urz) *noun* farmers who farm land for a landowner in return for a share of the crops
- **shantytown** (SHAN-tee-town) *noun* a community of shacks and other shelters built from scrap materials
- **socialism** (SOH-shuh-liz-uhm) *noun* an economic system in which the government, rather than private individuals, owns and operates factories, businesses, and farms
- **speculation** (spek-yuh-LAY-shuhn) *noun* the act of buying or selling stocks for a short term to profit from the rise or fall of prices
- **stockbrokers** (STAHK-broh-kurz) *noun* agents who buy or sell stocks or bonds for a client
- **veterans** (VET-ur-uhnz) *noun* persons who have served in the armed forces

FIND OUT MORE

BOOKS

Brown, Don. *The Great American Dust Bowl.* Boston: HMH Books for Young Readers, 2013.

Freedman, Russell. *The Children of the Great Depression.* Boston: HMH Books for Young Readers, 2010.

McDaniel, Melissa. *The Great Depression.* New York: Children's Press, 2012.

FILMS

Bound for Glory (1976). DVD, MGM, 2000.

The Grapes of Wrath (1940). DVD, 20th Century Fox, 2004.

I Was a Fugitive From a Chain Gang (1933). DVD, Warner Home Video, 2005.

Wild Boys of the Road (1933). In DVD collection, *Forbidden Hollywood: Volume Three*, Warner Archive Collection, 2009.

NOTE: Some books and films may not be appropriate for younger viewers.

VISIT THIS SCHOLASTIC WEBSITE
FOR MORE INFORMATION ABOUT
THE **GREAT DEPRESSION**

www.factsfornow.scholastic.com

Enter the keywords **GREAT DEPRESSION**

INDEX

African Americans, *82*, 84, *85*
art, *118*, 120, *121*

bankruptcies, 20, 76
banks, 8, 16, 20, *22*, 24, *25*, 60, *86*, 88, 135, 136, 137
Bethune, Mary McLeod, 84
Black Tuesday, 20, 136
Bonus Army, *34*, 36, *37*, 136

children, *42*, 44, *45*, 68, *69*, 92, *93*, *94*
Civilian Conservation Corps, 60
Congress, 36, *37*, 56, 60, 72, 128, 136, 137
Congress of Industrial Organizations (CIO), *114*, 116, 135, 138
Constitution of the United States, 72, 137
crime, 44, 72, *86*, 88, *89*, 135, *135*, 137

Democratic Party, 54, 56, 136
Dillinger, John, *86*, 88
Dust Bowl, *13*, *74*, 76, *77*, *78*, 80, 137

elections, 9, *54*, 56, 100, *101*, 104, 106, 108, *109*, 128, 134, 136, *136*, 138, 139
Empire State Building, 28, 124
employment, 8–9, 11, 16, 20, 28, *30*, 32, 36, 44, 48, 60, *61*, *62*, 64, *65*, 84, *85*, 96, 108, 116, *117*, 120, 128, *129*, 134, 136, 137

factories, 8, 10, 16, 116, *117*, 128, *129*
farming, 8, 10, 76, *77*, *78*, 84, *85*, 92, 108
Farm Security Administration (FSA), 92
Federal Art Project, *118*, 120, *121*
Federal Bureau of Investigation (FBI), 88
fireside chats, 68, *69*, 137, *137*
Floyd, Charles Arthur "Pretty Boy," 88, 135, *135*, 137
food, 32, *33*, 40, *42*, 44, *46*, 76, 84
Franco, Francisco, 96, 135, *135*, 138

General Motors, 116, *117*, 124
government assistance, 28, 32, 36, 108, 132
The Grapes of Wrath (John Steinbeck), *78*, 80, *81*
Great Plains, *74*, 76, 137

Hitler, Adolf, 96, 135, *135*
hoboes, *46*, 48, *49*, 52
Hoover Dam, 9, 64, *65*
Hoover, Herbert, 9, 24, 26, 28, *29*, 36, 40, *41*, *54*, 56, 64, 108, 134, *134*, 136
Hoovervilles, 40, *41*, *45*

inflation, 11, 96

labor unions, *114*, 116, *117*, 135, 138
Landon, Alf, 108, 134, *134*
Lewis, John L., *114*, 116, 135, *135*, 138, *138*
literature, *78*, 80
Long, Huey, 98, 100, *101*, 134, *134*, 138

maps, *12*, *13*
Medicare, 132
migrant workers, 48, 76, 80
movies, 52, 80, *81*, *110*, 112, *113*
music, 48, *50*, 52, *53*, 112, 120, *125*
Mussolini, Benito, 96, *97*

National Recovery Administration (NRA), 60
Nazi Party, *94*, 96, *97*
New Deal, 60, *61*, *62*, 64, *65*, 84, 108, *118*, 120, 128, 132, 134
New York Stock Exchange, *17*, 20
photography, *45*, 90, 92, *93*
Prohibition, *70*, 72, *73*, 88, 137, *137*

radio, 8, 52, *66*, 68, *69*, 100, 124, 137, *137*
recessions, 11
Reconstruction Finance Corporation, 24
Republican Party, 52, *54*, 56, 96, 104, 108, 128, 134, 135, 138
Roosevelt, Eleanor, 84
Roosevelt, Franklin Delano, *54*, 56, *57*, 58, 60, *66*, 68, *69*, 72, 84, 100, 106, 108, *109*, 124, 128, 134, *134*, 136, *136*, 137, *137*, 138, 139

savings accounts, 24, *25*, 104, 136
segregation, *82*, 84
shantytowns, 36, *38*, 40, *41*, *45*, 48
sharecroppers, 84, *85*, 92
sit-down strikes, 116, *117*
Social Security Act (1935), 104, *105*, 132, 138
stock market, 8, 10, *14*, 16, *17*, 20, *21*, 28, 104, *133*, 134, 136, *136*

taxes, 28, 100
television, 124
Tennessee Valley Authority (TVA), 64
theater, 80, 120

Wall Street, 8, *21*, *133*
Willkie, Wendell, 128
Works Progress Administration (WPA), 60, *61*, 80, *118*, 120, *121*
world's fair (1939), *122*, 124, *125*, 139
World War I, 36, 96, 136
World War II, 96, *126*, 128, *129*, 132, *133*, 135, 139, *139*

ABOUT THE AUTHOR

Steven Otfinoski has written more than 190 books for young readers. He is the author of several other volumes in the A Step Into History series, including *The Civil War* and *World War II*. Three of his books have been named to the New York Public Library's list of recommendations, Books for the Teen Age. He also teaches college English and creative writing. He lives with his wife in Connecticut.

Photos ©: cover: Fotosearch/Getty Images; 2: Roger-Viollet/The Image Works; 9: Bettmann/Getty Images; 10: The Granger Collection; 11: Bettmann/Getty Images; 14: Buyenlarge/Getty Images; 17: National Geographic Stock: Vintage Collection/The Granger Collection; 21: Bettmann/Getty Images; 22: Detroit Publishing Co./Library of Congress; 25: Bettmann/Getty Images; 29: John Tresilian/NY Daily News Archive/Getty Images; 30: Sueddeutsche Zeitung Photo/Alamy Images; 33: Everett Collection Inc./age fotostock; 34: Keystone-France/Getty Images; 37: MPI/Getty Images; 38: Roger-Viollet/The Image Works; 41: Image Asset Management/age fotostock; 42: Science History Images/Alamy Images; 45: Dorothea Lange/The Granger Collection; 46: American Stock Archive/Archive Photos/Getty Images; 49: Alan Fisher/Library of Congress; 50: Boyer/Roger Viollet/Getty Images; 53: Eric Schaal/The LIFE Picture Collection/Getty Images; 54: Hulton Archive/Getty Images; 57: Everett Collection Inc./age fotostock; 61: akg-images/The Image Works; 62: Margaret Bourke-White/The LIFE Picture Collection/Getty Images; 65: Dick Whittington Studio/Corbis/Getty Images; 66: George Skadding/The LIFE Picture Collection/Getty Images; 69: Bettmann/Getty Images; 70: CL Shebley/Shutterstock; 73: AP Images; 74: Bert Garai/Keystone/Hulton Archive/Getty Images; 77: UniversalImagesGroup/Getty Images; 78: Wrap-around colour book cover for the first edition of John Steinbeck's 'The Grapes of Wrath', 1939 (pictorial dust jacket, cloth slipcase), Hader, Elmer (1889-1973)/Private Collection/Photo © Christie's Images/Bridgeman Art Library; 81: Photo 12/Alamy Images; 82: The Granger Collection; 85: H. Armstrong Roberts/ClassicStock/Getty Images; 86: Topham/The Image Works; 89: GL Archive/Alamy Images; 93: Dorothea Lange/Library of Congress; 94: Bettmann/Getty Images; 97: Luce/Keystone/Getty Images; 101: ZUMAPRESS.com/age fotostock; 102: vkbhat/Getty Images; 105: AP Images; 109: Imagno/Austrian Archives/The Image Works; 110: SilverScreen/Alamy Images; 113: Everett Collection, Inc./Alamy Images; 114: Bettmann/Getty Images; 117: Tom Watson/NY Daily News Archive /Getty Images; 118: New York Times Co./Getty Images; 121: VCG Wilson/Corbis/Fine Art/Getty Images; 122: Library of Congress/Corbis/VCG/Getty Images; 125: Hulton Archive/Getty Images; 126: Museum of Flight/Corbis/Getty Images; 129: H. Armstrong Roberts/ClassicStock/Getty Images; 133: Bettmann/Getty Images; 134 center bottom: Bettmann/Getty Images; 134 bottom: Bettmann/Getty Images; 134 top: Hulton Archive/Getty Images; 134 center top: George Skadding/The LIFE Picture Collection/Getty Images; 135 top: Bettmann/Getty Images; 135 center top: Bettmann/Getty Images; 135 center bottom: Hugo Jaeger/Timepix/The LIFE Picture Collection/Getty Images; 135 bottom: Universal History Archive/UIG/Getty Images; 136 top: Buyenlarge/Getty Images; 136 bottom: Hulton Archive/Getty Images; 136 center: Sueddeutsche Zeitung Photo/Alamy Images; 137 top: CL Shebley/Shutterstock; 137 bottom: George Skadding/The LIFE Picture Collection/Getty Images; 138: Bettmann/Getty Images; 139 bottom: Museum of Flight/Corbis/Getty Images; 139 top: Bettmann/Getty Images.

Maps by Jim McMahon.